Jack and his mother had
each other, but not much
more. A year ago, a great big
beast took their things. Did
the mean beast rob them clean?

Jack's mother came to him
in tears.

"We have little to eat,"
she said. "Please trade the ox
for food."

2

A man with a beard bought
the ox. The man gave Jack
three yellow beans.

3

"That man is a cheat!"
Jack's mother said.

She tossed each bean over
her head.

Jack looked for each bean.
But there was only a great big
plant. And it was growing, up
to the sky.

With no fear, Jack went
up the bean plant. At the top
was a great big home.

The big mailbox read:

A mean beast lives here!

A mean beast lives here

Jack took back each
thing from the mean beast.

Then the mean beast
appeared!

"My ax, please, Mom,"
Jack screamed. "I need to
break the plant!"

The great plant fell, and
the mean beast was left high
overhead.

The End